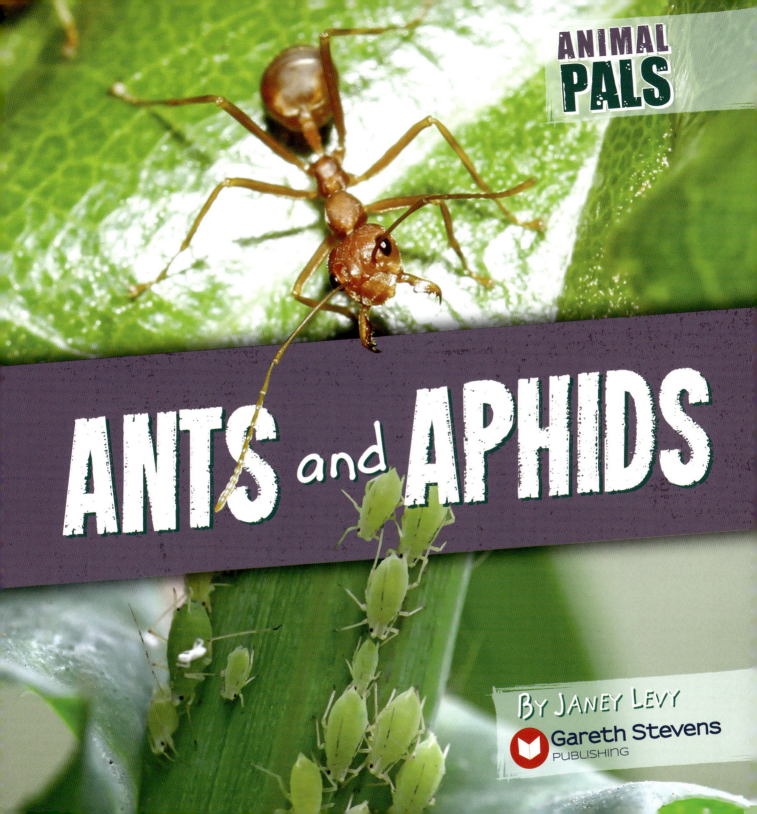

Please visit our website, www.garethstevens.com. For a free color catalog of all our high-quality books, call toll free 1-800-542-2595 or fax 1-877-542-2596.

Library of Congress Cataloging-in-Publication Data

Names: Levy, Janey, author.
Title: Ants and aphids / Janey Levy.
Description: New York : Gareth Stevens Publishing, [2022] | Series: Animal pals | Includes index.
Identifiers: LCCN 2020033170 (print) | LCCN 2020033171 (ebook) | ISBN 9781538266724 (library binding) | ISBN 9781538266700 (paperback) | ISBN 9781538266717 (set) | ISBN 9781538266731 (ebook)
Subjects: LCSH: Ants–Juvenile literature. | Aphids–Juvenile literature. | Mutualism (Biology)–Juvenile literature.
Classification: LCC QL568.F7 L48 2022 (print) | LCC QL568.F7 (ebook) | DDC 595.79/6–dc23
LC record available at https://lccn.loc.gov/2020033170
LC ebook record available at https://lccn.loc.gov/2020033171

First Edition

Published in 2022 by
Gareth Stevens Publishing
29 E. 21st Street
New York, NY 10010

Copyright © 2022 Gareth Stevens Publishing

Designer: Andrea Davison-Bartolotta
Editor: Monika Davies

Photo credits: Cover (top) PUMPZA/Shutterstock.com; cover (bottom) Inzyx/iStock/Getty Images Plus; p. 5 hekakoskinen/iStock/Getty Images Plus; p. 7 furryclown/Shutterstock.com; p. 8 Kletr/Shutterstock.com; p. 9 metel_m/Shutterstock.com; p. 11 Jaanus JSrva/Focus/Universal Images Group via Getty Images; p. 13 TrichopCMU/iStock/Getty Images Plus; p. 15 happystock/Shutterstock.com; p. 17 ValentynVolkov/iStock/Getty Images Plus; p. 19 Dimijana/iStock/Getty Images Plus; p. 20 Decha Thapanya/Shutterstock.com; p. 21 Creativ Studio Heinemann/Getty Images.

All rights reserved. No part of this book may be reproduced in any form without permission in writing from the publisher, except by a reviewer.

Printed in the United States of America

CPSIA compliance information: Batch #CSGS22: For further information contact Gareth Stevens, New York, New York at 1-800-542-2595.

CONTENTS

Better Together 4

Get Acquainted with Ants .6

More About Aphids 8

"Farmer" Ants and Aphid "Livestock" 10

How Ants Herd Aphids. .12

How Ants Milk Aphids 14

What's in It for Aphids? .16

How Ants Protect Aphids 18

Ants and Aphids Around the World 20

Glossary. 22

For More Information. .23

Index 24

Words in the glossary appear in **bold** type the first time they are used in the text.

BETTER TOGETHER

You've likely seen plenty of ants. Ants are everywhere, and they eat almost anything. But do you know what aphids are? They're tiny bugs that feed on plant sap. So, what do these two kinds of bugs have to do with each other? You might be surprised.

Ants and aphids often form a special **relationship** that has many benefits for both of them. This type of relationship is called mutualism. Inside this book, you'll learn more about how ants and aphids help each other.

FACT FINDER

Today, there are over 10,000 species, or kinds, of ants and over 4,000 species of aphids!

Ants and aphids have had a mutualistic relationship for millions of years!

GET ACQUAINTED WITH ANTS

What do ants look like up close? They have three body parts: a large head, a middle part called a thorax, and an abdomen, or stomach. A very small waist connects an ant's abdomen to its thorax. Ants also have a hard **exoskeleton** that covers their entire body.

Ants have six legs, two antennae, and two sets of strong **jaws**. Their outer jaws help them carry food and dig, and their inner jaws help them chew food.

FACT FINDER

Ants live in colonies led by a queen whose job is to lay eggs. Female worker ants hunt for food, care for the young, and **protect** the colony. Male ants **mate** with the queen.

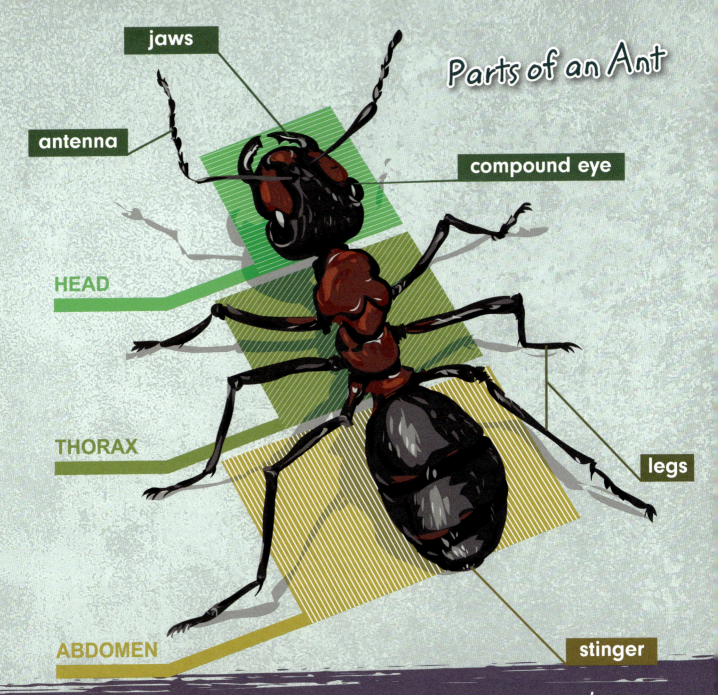

Ants also have compound eyes, which are eyes made of many separate seeing parts. Some kinds of ants have a stinger at the tip of their abdomen!

MORE ABOUT APHIDS

Aphids are so tiny you've likely never noticed them. Even the biggest aphids are only about 0.16 inch (4 mm) long! They have soft bodies that may be black, green, red, yellow, brown, or gray. Aphids have two antennae and long, sharp mouthparts that look a bit like thin tubes.

Aphids live in colonies on plant leaves and stems. In some places, colonies may have no males, because females can **reproduce** without mating! Females might lay eggs or give birth to live young called nymphs (NIMFS).

FACT FINDER!
Young aphids grow up fast. It takes them only about 7 to 10 days to go from baby to adult!

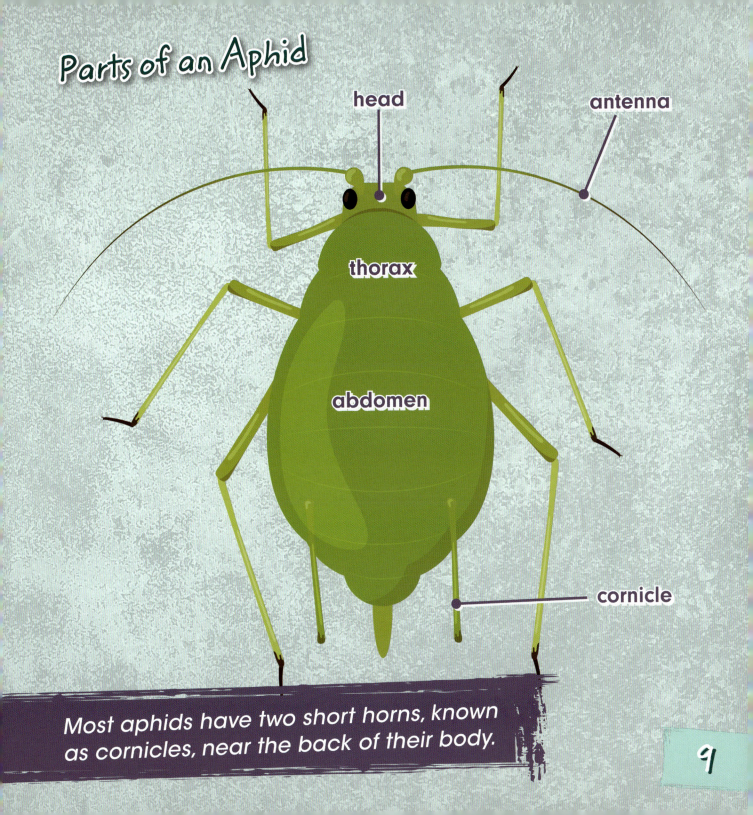

Most aphids have two short horns, known as cornicles, near the back of their body.

"FARMER" ANTS AND APHID "LIVESTOCK"

Ants and aphids are often paired up in the animal world. In this relationship, ants act as "farmers" that look out for aphids. Ants care for aphids much like a farmer might care for livestock. They keep aphids safe from predators and **shelter** them from bad weather.

Aphids produce a sweet, sugary liquid called honeydew. They make this from the plant sap they eat. Ants love drinking honeydew. The ants care for the aphids, and the aphids share the honeydew they make with ants.

Just like some farmers work hard to take care of their livestock, ant farmers work hard to take care of aphids.

FACT FINDER

Just like you, ants need more than sugar to live. Ants also need **protein**. Sometimes they eat their aphids.

HOW ANTS HERD APHIDS

Ants often herd aphids. This keeps aphids together in one area. But how do ants get a large number of aphids to stay in one place and not wander away to a new plant?

Ants have several tricks for herding aphids. They bite aphids' wings off so they can't fly away. Some ants produce **chemicals** that limit the growth of the aphids' wings. Their feet also create chemicals that make the aphids move slowly and calmly in the area the ants choose!

FACT FINDER

Just like human farmers might move livestock to a different field to get better food, ants may move the aphids to better feeding spots. The ants pick up aphids and carry them to the new location.

It's easier for ants to take care of an aphid herd when the aphids are all together in one place.

HOW ANTS MILK APHIDS

Dairy milk comes from cows. Human farmers must milk cows by hand or with a machine to get dairy milk. In a similar way, ants milk aphids to get the honeydew they want.

So, how do ants milk aphids? Ants don't have milking machines, and they don't have humanlike hands. Instead, ants stroke the aphids using their elbow-shaped antennae. This causes the aphids to create drops of honeydew. And the ants get a sweet drink of honeydew!

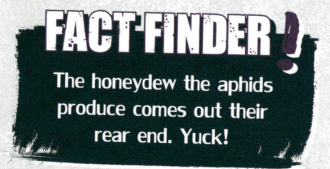

FACT FINDER
The honeydew the aphids produce comes out their rear end. Yuck!

For ants, honeydew is full of nutrients. Nutrients are what a living thing needs to grow and stay alive.

WHAT'S IN IT FOR APHIDS?

Ants benefit greatly from their relationship with aphids. But, remember, this is a mutualistic relationship between the two. So, how do aphids also benefit from their relationship with ants?

Aphids are very tiny and, unlike ants, have soft bodies. That means they face much danger in the world. If you were an aphid, it'd be a great help if someone was around to protect you when danger was near. And that's what ants do. They help keep aphids safe from enemies. Ants also help keep aphids well fed.

Farmers aren't big fans of aphids. Aphids often eat large parts of a plant to get the nutrients they need.

FACT FINDER!

Aphids move around very slowly. Their slow pace also puts them in danger from fast-moving predators.

HOW ANTS PROTECT APHIDS

Ants protect aphids in many ways. They attack bugs that eat aphids, such as ladybugs, spiders, and wasps. Some ants even destroy the eggs of aphid predators to protect aphids from **future** attacks!

Ants protect aphids from winter weather by taking their eggs into the ants' nests. In spring, after baby aphids **hatch**, ants carry them out of the nest and put them on plants to feed. Ants also protect their herds by removing sick aphids so the rest of the aphids don't get sick.

FACT FINDER

Some aphids have ways to protect themselves. If they're attacked by a predator, two chemicals in their body mix together and let out a deadly poison!

Aphids that aren't protected by ants are easy meals for predators such as ladybugs.

ANTS AND APHIDS AROUND THE WORLD

The mutualistic relationship between ants and aphids provides great benefits for both bugs. It works so well that you can find them paired up around the world.

Ants and aphids are found together on every **continent** except Antarctica. It's too cold for the two bugs to survive in Antarctica. But you can find ants and aphids paired up in North America, South America, Europe, Asia, Africa, and Australia. You might even find them in your own backyard or in a nearby park!

Ants and aphids have built a life together that helps both of them survive.

FACT FINDER!

While some ants and aphids are often found aboveground, others are hidden away. Some ants and aphids live and work together underground, deep in the soil!

GLOSSARY

chemical: matter that can be mixed with other matter to cause changes

continent: one of Earth's seven great landmasses

exoskeleton: the hard outer covering of an animal's body

future: the period of time that will follow or occur after the present time

hatch: to break open or come out of

jaws: the walls of the mouth

mate: to come together to make babies

protect: to keep safe

protein: a nutrient in many types of food that the body uses to grow, repair tissues, and stay healthy

relationship: a connection between two living things

reproduce: when an animal creates another creature just like itself

shelter: to keep animals or people safe

FOR MORE INFORMATION

Books

Kalman, Bobbie. *Symbiosis: How Different Animals Relate.* New York, NY: Crabtree Publishing, 2016.

Morlock, Rachael. *Ants Up Close.* New York, NY: PowerKids Press, 2019.

Perish, Patrick. *Aphids.* Minneapolis, MN: Bellwether Media, 2018.

Websites

Ants Farming Aphids!
www.nts.org.uk/stories/ants-farming-aphids
Read more about ants farming aphids, and watch a short video on this site.

Ants, Fourmis
www.biokids.umich.edu/critters/Formicidae/
Are you curious about ants? Discover more about their eating and life habits here.

Ant vs Ladybird
www.bbc.co.uk/programmes/p003lckg
Watch a quick video of ants protecting their herd of aphids from a ladybird, or ladybug.

Publisher's note to educators and parents: Our editors have carefully reviewed these websites to ensure that they are suitable for students. Many websites change frequently, however, and we cannot guarantee that a site's future contents will continue to meet our high standards of quality and educational value. Be advised that students should be closely supervised whenever they access the internet.

INDEX

abdomen, 6, 7, 9

antennae, 6, 7, 8, 9, 14

chemicals, 12, 18

colony, 6, 8

compound eyes, 7

continents, 20

cornicles, 9

eggs, 8, 18

exoskeleton, 6

honeydew, 10, 14, 15

jaws, 6, 7

ladybugs, 18, 19

legs, 6, 7

mutualism, 4

nutrients, 15, 17

nymphs, 8

predators, 10, 17, 18, 19

protein, 11

queen, 6

sap, 4, 10

spiders, 18

stinger, 7

thorax, 6, 7, 9

wasps, 18

wings, 12